Progressive Math

A curriculum that links existing knowledge to new content

Level 1

By: Dr. Lynne Gregorio

www.apex-math.com

Copyright © 2010 by Progressive Educational Press

Printed in the United States of America

v. 1.0

ISBN: 978-0-9828903-0-1

This book is available at special discounts for bulk purchases.
For more information contact the author at
lynne@apexlearningcenter.com

_____

Editor: Kelsey Carter

Table of Contents

About this book

This book is designed with the concept that students learn better and retain more if they link existing knowledge to new content they are learning. Students will be more successful and become less frustrated with math if they do smaller steps leading up to a new concept rather than taking bigger conceptual leaps. We feel it is also important to always review older concepts along the way to learning newer concepts. This allows for students to move at a faster pace as they don't have to do too many problems for each concept and they don't have to go through a re-teaching period for the concepts they have forgotten.

Acknowledgments

I would like to thank my husband, Joe, for always being supportive of my desire to teach and help people understand and for all the technical support he has provided to my projects. I would also like to thank my four children, Christopher, Austin, Reilly, and Caden for always letting mom try new math ideas out with them to find the optimal way to teach a subject. Finally, I would like to say thank you to my Apex High School AOIT intern, Kelsey, for all her hard work that went into this project.

On the subject of Scope and Sequence

In order for any mathematics curriculum to be successful, you need to start with a list of goals. There are many topics in mathematics that could be included in a math curriculum. A decision must be made on what topics to cover.

We have chosen specific goals. First, we want our goals to be linked to current knowledge and what the student is learning. We don't let a traditional concept of sequencing dictate our choice of order. Instead, we let the idea of the concept itself dictate the order. For example, we spend a good part of level 1 teaching telling time and counting coins. The reason for this is because the concept "counting by 5's" is linked to telling time and counting coins, so it is reasonable to apply those concepts when talking about counting by 5's.

We have chosen to keep our focus on measurement (counting coins and telling time count as a form of measurement), arithmetic, and problem solving. We feel that learning a large number of Geometry vocabulary words (as is taught in the traditional curriculum) is not worthwhile. We will reach a point when we do an entire unit on Geometry or we may do Geometry ideas as an application of addition or multiplication but have purposely chosen not to have students memorize words such as "acute angle" and "obtuse angle" when there are more practical ways to use the student's educational time. This level 1 curriculum introduces the topics listed.

Scope and Sequence

- Counting by 5's and 10's.
- Telling time to the 5 minute interval
- Counting up to 3 coins at a time
- Addition facts
- Subtraction facts
- Number sense
- Place value to the hundreds place
- Applications / problem solving
- Multiplication X2
- Division (cut smaller numbers in half)

Age, Grades, and Levels

We also feel that age and grade should not dictate or limit a student's progression in mathematics. Some students will move quite quickly at a young age through the levels. For others, it may take more time and practice. We say that level 1 is generally applicable for grades K-2. Although advanced pre-school children may easily be able to begin the curriculum. We recommend that students can read as we want them to be able to figure out the word problems on their own by applying the process of looking for key words and reading comprehension but the curriculum can be done with the parent reading the problems to the student if needed. We also say that this curriculum can apply to students as old as 2nd grade because many of them can only "count on their fingers" instead of having a strong number sense. They also may not have learned good strategies for counting coins and telling time or may still be struggling with knowing when to add or subtract in word problems, so this curriculum would be a perfect way to improve those important skills even if the student is beginning to move on to dynamic addition (addition with re-grouping).

Lesson 1: Counting by 10's

Goal: **To learn to count by 10's in a variety of formats.**

New Teaching: Children easily learn to count by 10's in kindergarten. In this lesson, we take it a step further and teach them to apply the process of counting by 10's to multiple situations. We introduce the dime at this time because it is linked to counting by 10's. We also introduce more challenging patterns using the same concept that will help them later on and encourage critical thinking. Make sure the child sees and understands these patterns.

Student Practice

Complete the pattern by counting by 10's.

1. 10, 20, 30, _____, _____, _____

2. 30, 40, 50, 60, _____, _____, _____

3. 10, _____, 30, _____, 50, _____, _____

4. 40, 50, 60, _____, _____, _____

5. 25, 35, 45, _____, _____, 75, _____, _____

6. 12, 22, 32, 42, 52, _____, _____, _____

7. 46, 56, 66, 76, _____, _____, _____

8. 45, 55, 65, _____, _____, 95

9. 19, 29, 39, 49, _____, _____, _____, 89

10. 13, 23, 33, 43, _____, _____, _____

11. 21, 31, _____, _____, 61, _____, _____, 91

12. 44, 54, _____, _____, _____, 94

13. 37, 47, _____, _____, _____, _____

14. 2, 12, _____, _____, 42, _____, _____

15. Count the dimes (each dime has a value of 10 cents).

_____ cents

16. Count the dimes (each dime has a value of 10 cents).

_____ cents

17. Count the dimes (each dime has a value of 10 cents).

_____ cents

18. Count each X as 10. Note that in **Roman Numerals,** the X stands for 10. X X X X X   _____

19. Count each X as 10. Note that in **Roman Numerals,** the X stands for 10. X X X X X X X   _____

Notes:

_____

_____

_____

_____

_____

_____

_____

_____

_____

_____

_____

_____

Lesson 2: Ordering Numbers

<u>Goal:</u> **To understand the comparative order of numbers.**

<u>New Teaching:</u> Number sense is essential to building strong mathematics skills. This lesson has the student practice finding values that come before and after given numbers. This will help them later when they rely on that information for addition and subtraction.

In some of the problems you will see the word written out. These may be more challenging for some students. They may need your help spelling the word, but have the student enter the number in the same form it is presented.

Student Practice

What comes in the middle?

1. 4, _____, 6

2. 8, _____, 10

3. 10, _____,12

4. 14, _____, 16

5. Four, _____, Six

6. Nine, _____, Eleven

7. Fifteen, _____, Seventeen

8. 35, _____, 37

9. 51, _____, 53

10. 73, _____, 75

11. 96, _____, 98

Student Practice

What comes after?

1. 3, _____

2. 6, _____

3. 9, _____

4. 13, _____

5. 17, _____

6. 24, _____

7. 56, _____

8. 89, _____

9. Twelve, _____

10. Forty, _____

11. Sixty-five, _____

Student Practice

What comes before?

1. _____, 5

2. _____, 9

3. _____, 14

4. _____, 45

5. _____, 32

6. _____, 81

7. _____, 29

8. _____, 40

9. _____, 60

10. _____, Ten

11. _____, Nineteen

Student Practice

Count by 10's.

1. 50, 60, 70, _____, 90

2. 35, 45, 55, _____, _____, _____

3. 10, 20, _____, 40, _____, _____

4. 13, 23, 33, 43, _____, _____, _____

5. _____, 24, 34, _____, _____, _____, 74

6. _____, 46, 56, _____, 76, _____, _____

7. Count the dimes (each dime is worth 10 cents).

    _____ cents

8. Count each X as 10. Note that in **Roman Numerals** the X

   stands for 10. X X X X _____

Notes:

_____

_____

_____

_____

_____

_____

_____

_____

_____

_____

_____

_____

_____

Lesson 3: Finger Sense

<u>Goal:</u> **To use fingers as a visual aid to see number patterns with 5 as an addend.**

<u>New Teaching:</u> Many children use their fingers to help them do mathematics early on. Our goal is to teach them to take it a step further and build additional number sense with their fingers. The goal of this unit is to link the numbers 6 – 10 with the value of 5 plus another number. We want the student to hear 7 and immediately think 5 and 2. By using fingers as visuals, children commit into their memory the concept that 7 is 5 + 2. We want to help students find ways to think about numbers as groups instead of individual units that must always be counted by ones.

This is a great exercise that parents can do while waiting for dinner at a restaurant or at a doctor's appointment or any of those other times we spend just waiting with our kids. Hold up fingers of 5 and 3 and ask the child to quickly tell you how many without counting! You can also reverse it and say, "Make 6 with your fingers, quickly, without counting."

From there, you can move into saying a number such as 7 and have the student tell you (without making the fingers) that 7 is 5 and 2.

How Many Fingers?

It is very helpful to use your fingers as you begin to learn math. One thing you need to know is how many fingers make certain numbers.

<u>Example:</u> To make 6, I need 1 hand and 1 finger.

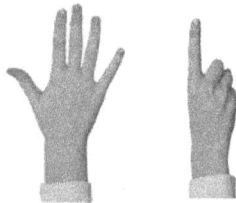

<u>Example:</u> To make 8, I need 1 hand and 3 fingers.

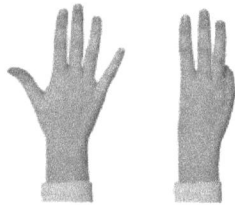

<u>Example:</u> To make 10, I need 2 hands.

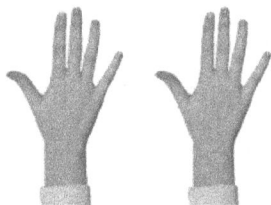

Student Practice

First, create the numbers using your hands. Then write down the amount of numbers on each hand that makes the total.

1. Make the number 7 with your hands. ____ and ____

2. Make the number 9 with your hands. ____ and ____

3. Make the number 6 with your hands. ____ and ____

4. Make the number 8 with your hands. ____ and ____

5. Make the number 10 with your hands. ____ and ____

Now, try to do it by "thinking" of it in your head, but without using your hands.

1. 6 is 1 hand and ____ fingers.

2. 9 is 1 hand and ____ fingers.

3. 8 is 1 hand and ____ fingers.

4. 7 is 1 hand and ____ fingers.

Student Practice

Remember that 1 hand means 5 fingers: 6 is 5 fingers + 1 finger or 5 + 1.

1. 7 is _____ fingers + _____ fingers or _____ + _____

2. 9 is _____ fingers + _____ fingers or _____ + _____

3. 6 is _____ fingers + _____ fingers or _____ + _____

4. 8 is _____ fingers + _____ fingers or _____ + _____

5. 10 is _____ fingers + _____ fingers or _____ + _____

Now, we will go the other way (try to do it without looking at your hands if you can).

1. 1 hand and 3 fingers is _____

2. 1 hand and 1 finger is _____

3. 1 hand and 5 fingers is _____

4. 1 hand and 2 fingers is _____

5. 1 hand and 4 fingers is _____

Student Practice

It's the same idea (1 hand = 5 fingers), only you are going to use numbers this time.

1. $7 = 5 +$ _____

2. $5 + 3 =$ _____

3. $6 = 5 +$ _____

4. $5 + 5 =$ _____

5. $9 = 5 +$ _____

6. $8 = 5 +$ _____

7. $5 + 4 =$ _____

8. $5 + 1 =$ _____

9. $10 = 5 +$ _____

10. $5 + 2 =$ _____

Mixed Review

1. 40, 50, 60, _____, _____, _____

2. 18, 28, _____, _____, _____, 68, _____, 88

3. Count the coins (each dime is worth 10 cents)

    _____ cents

4. 8 is made up of 1 hand and _____ fingers

5. 7 is made up of 1 hand and _____ fingers

6. 6 = 5 + _____

7. What comes before 29? _____

8. What comes after 49? _____

9. What comes before 80? _____

10. What comes after 51? _____

11. _____ comes before 76.

12. _____ comes after 38.

13. Twelve, _____, Fourteen

14. 56 comes before_____

15. 8 = 5 + _____

16. 5 + 2 = _____

17. 5 + 1 = _____

18. 9 = 5 + _____

19. Count the coins (each dime is worth 10 cents)

_____ cents

20. Count the coins (each dime is worth 10 cents)

_____ cents

21. 10, _____, _____, 40, _____

22. 57, 67, _____, _____, _____

23. 26, 36, _____, _____, _____

Notes:

_____

_____

_____

_____

_____

_____

_____

_____

_____

_____

_____

_____

_____

Lesson 4: Making Teens

Goal: **To understand the patterns in numbers when making teens.**

New Teaching: This is extremely important as students build number sense and begin to move towards addition and subtraction. It is great to start with making teens because it is an easy concept and most students pick it up quickly.

Lesson Dialogue: "As we begin to explore numbers, we note that some numbers are one-digit numbers, like 1, 4, and 8. Some numbers have 2 digits, such as 14, 32, and 85. The first digit in a two-digit number is called the tens place, as it tells us how many tens we have."

"Right now, we are going to work with our 'teen' numbers: 11, 12, 13, 14, 15, 16, 17, 18, and 19. All of these numbers are after the number 10. Therefore, they are bigger than 10. In fact, the first digit tells us we have 1 ten and the second digit tells us how many above 10 the number is."

<u>Example:</u> 13 is 10 plus 3 more. Imagine 2 full hands and 3 more fingers on someone else's hands.

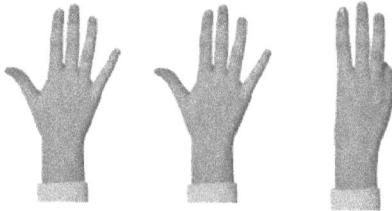

"So, 13 is 10 fingers (2 hands) and 3 more fingers (your friend's hand). 13 = 10 + 3."

"How many tens are in 14?" The student replies, "1."

"Good. How many ones are in 14?" The student replies, "4."

"How many tens are in 16?" The student replies, "1."

"Good. How many ones are in 16?" The student replies, "6."

"How many tens are in 27?" The student replies, "2."

"Good. How many ones are in 27?" The student replies, "7."

Student Practice

1. $13 = 10 +$ _____

2. $17 = 10 +$ _____

3. $11 = 10 +$ _____

4. $14 = 10 +$ _____

5. $19 = 10 +$ _____

6. $16 = 10 +$ _____

7. $12 = 10 +$ _____

8. $15 = 10 +$ _____

9. $18 = 10 +$ _____

10. $10 + 3 =$ _____

11. $10 + 4 =$ _____

12. $10 + 9 =$ _____

13. 10 + 1 = ____

14. 10 + 5 = ____

15. 10 + 2 = ____

16. 10 + 7 = ____

17. 10 + 6 = ____

18. 10 + 8 = ____

Were you able to see the pattern?

Lesson 5: Doubling 2, 3, and 4

Goals: **To learn the doubled values of 2, 3, and 4 and different ways to think about doubling a number.**

New Teaching: When learning addition facts and later multiplication facts, it is helpful if the student can easily double numbers. These require some memorization. Make some flash cards and have the student quiz themselves by using the flashcards. Spend some time with the child by quizzing them as well. We will double numbers three at a time so that it is not too much to do in one lesson. Mastery of doubling, however, is required to be able to apply future addition strategies that will be taught.

<u>Lesson Dialogue:</u> "Doubling means making 2 groups of the number and finding the total."

"Doubling two would look like this:" ●● ●●

"You can see if you have two groups of two, you have a total of 4 dots. Doubling two can be written as: 2 + 2 or 2 X 2. Both give 4 as an answer."

"Now let's double three:" ●●● ●●●

"We count and see that doubling 3 equals 6 dots. We can write this as 3 + 3 or as 2 X 3. When we write it as 2 X 3 – this means 2 groups of 3. The X sign means groups of."

Draw the picture to double four.

Make 2 groups of four. Count the number of dots total. _____

Write the math sentences:

_____ + _____ = 8

  OR

_____ groups of _____ = 8

  OR

_____ X _____ = 8

Student Practice

1. Double 2 = _____

2. Double 3 = _____

3. Double 4 = _____

4. 2 + 2 = 2 X _____

5. 3 + 3 = 2 X _____

6. 4 + 4 = 2 X _____

7. 2 + 2 = _____

8. 4 + 4=_____

9. 3 + 3=_____

10. 2 X 2 = _____

11. 2 X 3 = _____

12. 2 X 4 = _____

"Just as we learned to double a number, we can also do the opposite: divide the number in half."

Example:

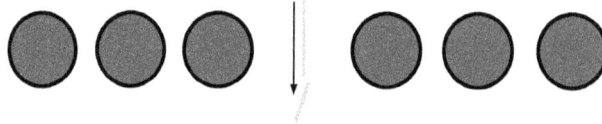

"We can see that by splitting the dots into two equal groups we get 3 in each group. Think of it this way. If you have 6 pieces of candy and ate half of them, you would now only have 3 left."

"For the numbers 2, 4, 6, and 8, draw the circles and find the line that splits them in half. Count the dots on one half and report this number."

Student Practice

1. Half of 2 _____

2. Half of 4 _____

3. Half of 6 _____

4. Half of 8 _____

Mixed Review

1. $5 +$ _____ $= 7$ (remember your fingers)

2. $8 = 5 +$ _____

3. _____ $= 5 + 4$

4. Count by 10's: 35, 45, 55, _____, _____, _____

5. Count by 10's: 14, 24, 34, _____, _____, _____

6. Count by 10's: 27, 37, _____, _____, _____, _____

7. What comes before 30? _____

8. What comes after 59? _____

9. What comes before 67? _____

10. 42, _____, 44

11. 96, _____, _____, 99

12. What comes after 80? _____

13. What is 4 + 4 (double 4)? _____

14. What is 2 X 4 (two groups of four)? _____

15. What is half of 4? _____

16. 14 = 10 + _____

17. 10 + 7 = _____

18. 19 = 10 + _____

19. 10 + 8 = _____

20. 10 + _____ = 15

21. Double 3? _____

22. Take half of 6? _____

23. Double 2? _____

24. 2 groups of 4? _____

25. 3 + 3 = 2 X _____

Notes:

_____

_____

_____

_____

_____

_____

_____

_____

_____

_____

_____

_____

_____

Lesson 6: Counting by 5's

Goal: **To learn to count by 5's using different approaches.**

New Teaching: Just as we learned to count by 10's, we need to learn how to count by 5's. This skill is very helpful in many different ways: counting nickels, tallies, telling time, and multiplication. We start off easy and then progress onto harder questions as well as different visuals that require counting by 5.

Draw a clock and have only the big hand point to a number. Have the student start at 1 and count by 5's until they get to the number that the hand points to. This is the first step in learning to tell time!

Lesson Dialogue: "We are going to learn how to count the minute hand on the clock. The minute hand is the biggest hand on the clock. When the big hand starts on the 12, it represents 0 minutes. When that hand is on the 1, 5 minutes have passed. So as the big hand moves around the clock, you are counting by 5's. Let's say the big hand is on the 4. You will count by 5's four times. 5, 10, 15, 20. Now let's say the hand is on the 2. How many minutes is that?"

The student responds, "You count by 5's twice. 5, 10. 10 minutes."

"Good. I'm going to give you some practice with a clock."

Example:

"What number is the hand on?" The student responds, "3."

"So how many times are you going to count by 5?" The student responds, "3 times."

"Good. So when the hand is on the 3, how many minutes have passed?" The student says, "5, 10, 15. 15 minutes."

"Excellent. Let's try another one."

Example:

"Where is the minute hand?" The student says, "It's on the 5. So I am going to count by 5's five times."

"Good. How many minutes have passed if the minute hand is on the 5?" The student counts, "5, 10, 15, 20, 25. 25 minutes."

"Right. Here's another one."

Example:

"What number is the minute hand on?" The student replies, "6. I will count by 5's six times."

"Correct."

The student says, "That is 5, 10, 15, 20, 25, 30. So when the hand is on the 6, 30 minutes have passed."

"Excellent! I have one more practice for you to do before you do this yourself."

Example:

"Tell me what number the hand is on."

The student answers, "The hand is on the 8. I am going to count by 5's eight times. 5, 10, 15, 20, 25, 30, 35, 40. So when the hand is on the 8, 40 minutes have passed."

"Correct! Now you will be practicing counting by 5's on your own."

Student Practice

Count by 5's.

1. 5, 10, 15, 20, _____, _____, _____, 40, 45

2. 15, 20, 25, 30, 35, _____, _____, _____, 55, 60

3. 55, 60, 65, 70, _____, _____, _____, 90

4. 40, 45, 50, 55, 60, _____, _____, _____, _____, _____

5. Count the value of the nickels. A nickel is worth 5 cents.

     Value _____

6. Count the value of the nickels. A nickel is worth 5 cents.

     Value _____

7. Count the number of tallies (count by 5's).

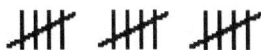

    卌 卌 卌   Number of Tallies _____

8. Count the number of fingers (count by fives).

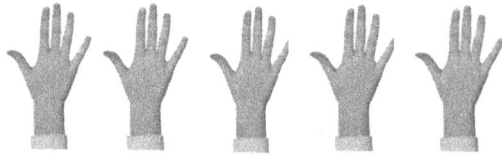

Number of Fingers _____

9. Count the hands on a clock. Each number represents 5.

So **for 1: say 5; for 2: say 10; for 3: say 15.** What

do you get if the hand is on the 4? _____

10. What do you get if the hand is on the 6? (Count by 5's

until you reach 6) _____

11. What do you get if the hand is on the 8? _____

12. What do you get if the hand is on the 9? _____

13. What do you get if the hand is on the 11? _____

Notes:

_____

_____

_____

_____

_____

_____

_____

_____

_____

_____

_____

_____

_____

Lesson 7: Place Value

<u>Goal:</u> **To learn the names for place value of ones, tens, and hundreds.**

<u>New Teaching:</u> Place value is vital to future mathematical study. Many of the concepts taught rely on the understanding and labeling of place value. Continue to work on this often, and make sure your student clearly understands the names that go with the different places. At this time, it is only important that they learn to the hundreds place. However, feel free to extend to the thousands place if you feel your child is ready and wants to learn. Kids often find it fun to look at a number in the millions place, so feel free to explore millions as well.

Lesson Dialogue: "There is a name and value associated with the location of a number. We will focus on three places or numbers in the hundreds."

Example: "The number 432 is read as four-hundred thirty-two. The 4 tells us how many hundreds, the 3 tells us how many tens (3 tens is 30) and the 2 is the number of ones."

"Make a chart that reminds you of the place value of each number."

Hundreds	Tens	Ones
4	3	2

Example: 975

Hundreds	Tens	Ones
9	7	5

Example: 26

Hundreds	Tens	Ones
0	2	6

Student Practice

Fill in the chart for each number.

Number	Hundreds	Tens	Ones
583			
901			
8			
35			
275			
298			
764			
420			
100			
315			

Student Practice

Tell the place value of each number (hundreds, tens, or ones).

1. The 5 in 503   _____

2. The 1 in 421   _____

3. The 8 in 834   _____

4. The 4 in 140   _____

5. The 0 in 350   _____

6. The 6 in 463   _____

7. The 9 in 934   _____

8. The 3 in 385   _____

9. The 2 in 326   _____

10. The 7 in 174   _____

11. The 6 in 296   _____

Student Practice

Example: 356 = 3 hundreds, 5 tens, and 6 ones

Write each number as shown above.

1. 465 _____

2. 619 _____

3. 462 _____

4. 874 _____

5. 903 _____

6. 274 _____

7. 555 _____

8. 130 _____

9. 782 _____

10. 345 _____

11. 206 _____

Student Practice

Expanded notation is another way to write the numbers that shows their place value.

<u>Example:</u> 659 = 600 + 50 + 9

<u>Example:</u> 204 = 200 + 4 (Note I didn't put 10's in because it had a 0 in the tens place)."

Write each number in expanded notation.

1. 465 _____

2. 219 _____

3. 874 _____

4. 903 _____

5. 274 _____

6. 555 _____

7. 130 _____

8. 782 _____

Lesson 8: Doubling 5, 6, and 7

Goals: To learn to double 5, 6, and 7 and continue to learn different ways of asking to double or do the inverse and take half.

New Teaching: Remember when we learned to double 2, 3, and 4. Now, we're going to double 5, 6, and 7.

Double 5: We get 10. When doubling 5, think of two hands. Two groups of 5 give you a total of 10 fingers.

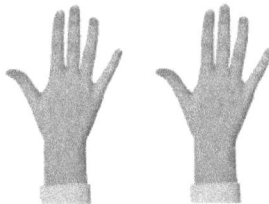

Double 6: Six is one **HALF-DOZEN** eggs. Two half-dozens make a whole dozen – **ONE DOZEN** eggs equals 12.

Half Dozen Eggs

One Dozen Eggs

Double 7: This is a harder one to remember. I use a deep voice when I say "7 + 7 = 14." I keep saying it in my deep voice and that often helps students remember this double!

What else should I remember?

- Doubling is the same as adding a number to itself: Double 5 means 5 + 5.

- Doubling is the same as multiplying that number by 2 (X2): Double 5 means 2 X 5. Remember the X means groups of – so 2 groups of 5 equals 10.

- If I have the doubled value such as 10, I can easily take half and get to 5.

Student Practice

1. Double 5 (2 hands)   ____

2. Double 6 (makes a dozen eggs) ____

3. Double 7 (say it in a deep voice) ____

4. $5 + 5 =$ ____

5. $7 + 7 =$ ____

6. $6 + 6 =$ ____

7. 2 "groups of" 5 = ____

8. $2 \times 5 =$ ____

9. 2 "groups of" 6 = ____

10. $2 \times 6 =$ ____

11. 2 "groups of" 7 = ____

12. $2 \times 7 =$ ____

13. 4 + 4 = _____

14. 3 + 3 = _____

15. 2 + 2 = _____

16. 2 X 3 = _____

17. 2 X 4 = _____

18. 2 X 2 = _____

19. Double 2 _____

20. Double 4 _____

21. Double 3 _____

22. Take HALF of 8 _____

23. Take HALF of 10 _____

24. Take HALF of 6 _____

25. Take HALF of 4 _____

Mixed Review

1. What comes after 49? _____

2. What comes before 30? _____

3. How many fingers make 7? 5 + _____

4. 13 = 10 + _____

5. 19 = 10 + _____

6. 10 + 6 = _____

7. 10 + 4 = _____

8. Using the hands on a clock, how many minutes have passed (count by 5's) if the big hand is on the 2? _____

9. Using the hands on a clock, how many minutes have passed if the big hand is on the 7? _____

10. Write 476 in expanded notation. _____

11. What is the place value of the 7 in 476?

12. How many hundreds are in 476? _____

13. How many tens are in 192? _____

14. What is the place value of the 2 in 192? _____

15. Double 4 _____

16. 5 + 5 = _____

17. Double 6 _____

18. 7 + 7 = ____

19. 2 X 4 = ____

20. 2 groups of 7 = ____

21. 2 groups of 5 = ____

22. 2 groups of 2 = ____

23. 2 X 3 = ____

24. 2 X 6 = ____

25. 2 X 5 = ____

26. What is half of 14? ____

27. What is half of 10? ____

28. What is half of 12? ____

29. What is half of 8? ____

30. What is half of 6? ____

Notes:

_____

_____

_____

_____

_____

_____

_____

_____

_____

_____

_____

_____

_____

Lesson 9: Count by 25's & Quarters

Goals: To learn how to count by 25's and apply to counting quarters.

New Teaching: Quarters are worth 25 cents. They are one-quarter of a dollar and this is how they get their name.

- 1 Quarter = 25 cents

- 2 Quarters = 50 cents (this is ½ of a dollar)

- 3 Quarters = 75 cents

- 4 Quarters = $1 dollar (this is also 100 cents!)

When counting by 25's, we would say, "25, 50, 75, $1."
(Hint: Do this in a sing-song voice to help with memory)

Student Practice

Count by 25's to solve these problems.

1. 25, 50, _____, $1

2. 25, _____, 75, $1

3. 1 Quarter = _____ cents

4. 2 Quarters = _____ cents

5. 3 Quarters = _____ cents

6. 4 Quarters = $_____

7. How many quarters are in 50 cents? _____

8. How many quarters make 25 cents? _____

9. How many quarters are in 1 dollar? _____

10. 25 + 25 = _____

11. 25 + 25 + 25 = _____

12. 4 groups of 25 = _____

13. 3 groups of 25 = _____

14. How many cents? _____

15. How many cents? _____

16. How many cents? _____

17. How much money? _____

Notes:

_____

_____

_____

_____

_____

_____

_____

_____

_____

_____

_____

_____

Lesson 10: Counting One Type of Coin

Goals: **To reinforce how many cents are in a nickel, dime, and quarter and practice counting single coins.**

Money

We have learned that:

- A nickel is 5 cents.

- A dime is 10 cents.

- A quarter is 25 cents.

- You may know that a penny is 1 cent.

Let's practice our counting. Remember when counting nickels, count by 5's and when counting dimes, count by 10's. When counting quarters, count by 25's and when counting pennies, count by ones.

Student Practice

Count the coins to find the total number of cents.

1. _____ cents

2. _____ cents

3. _____ cents

4. _____ cents

5. _____ cents

6. _____ cents

7. _____ cents

8. _____ cents

9. _____ cents

10. _____ cents

11. _____ cents

Notes:

_____

_____

_____

_____

_____

_____

_____

_____

_____

_____

_____

_____

_____

Lesson 11: Counting Nickels and Pennies

<u>Goal:</u> **To be able to count groups of nickels and pennies by switching between counting by 5's and counting by 1's.**

Counting Nickels and Pennies Together

When counting by nickels and pennies, we have to count by 5's and then switch to counting by 1's.

<u>Example:</u> 5, 10, 15, 20 SWITCH 21, 22, 23

<u>Example:</u> 5, 10, 15, 20, 25 SWITCH 26, 27, 28

Student Practice

Count by 5's and then count by 1's after the word switch.

1. 5, 10, 15, 20 SWITCH 21, 22, 23, _____, _____

2. 5, 10, 15 SWITCH 16, 17, 18, _____

3. 5, 10, 15, 20, 25, 30 SWITCH 31, 32, _____, _____

4. 5, 10 SWITCH 11, 12, _____, _____

5. 5, 10, 15, 20 SWITCH _____, _____, _____

6. 5, 10, 15, _____, _____, _____ SWITCH 31, 32, _____, _____

7. 5, 10, _____, _____, _____ SWITCH _____, _____, _____

8. 5, 10, 15, 20, 25, _____, _____, _____ SWITCH _____, _____

9. 5, 10, 15, _____, _____ SWITCH _____, _____, _____, _____

10. 5 SWITCH _____, _____

11. 5, 10, _____ SWITCH _____, _____, _____

12. 5, 10, _____, _____, _____, _____, _____ SWITCH _____, _____

13.

_____ _____ _____ SWITCH _____ _____

14.

_____ _____ _____ _____ _____ SWITCH _____

15.  _____ cents

16.  _____ cents

17.  _____ cents

Mixed Review

1. 19, 29, 39, ____, ____, ____, ____.

2. 84 is before ____ and after ____

3. ____ comes after 24.

4. ____ is before 63.

5. 9 = 5 + ____

6. 5 + 2 = ____

7. 4 X 2 = ____

8. Double 7? ____

9. Half of 12? ____

10. Half of 6? ____

11. 25, ____, 75, $1.

12. 3 groups of 25? ____

13. _____ cents

14. _____ cents

15. _____ cents

16. _____ cents

17. _____ cents

18. 5, 10, 15, _____, _____, _____

19. 65, 70, 75, _____, _____, _____, _____

20. How many dimes are in 20 cents? _____

21. How many nickels are in 35 cents? _____

Notes:

_____

_____

_____

_____

_____

_____

_____

_____

_____

_____

_____

_____

Lesson 12: Adding + 1

<u>Goal:</u> **To learn to add a number + 1 by counting on to the next number.**

<u>New Teaching:</u> When adding 1 to a number, we want 1 higher than the current number. That is the same as saying "what number comes next?" Let's add one to each of the following:

- 3 + 1 = 4

- 5 + 1 = 6

- 14 + 1 = 15

- 23 + 1 = 24

- 78 + 1 = 79

Remember, it is just the number that comes next because you are making it one bigger. It is important to make sure your child is not counting when doing these problems. They need to understand that they should just "find the next number" when adding one.

Student Practice

1. $3 + 1 =$ _____

2. $9 + 1 =$ _____

3. $5 + 1 =$ _____

4. $8 + 1 =$ _____

5. $10 + 1 =$ _____

6. $15 + 1 =$ _____

7. $23 + 1 =$ _____

8. $55 + 1 =$ _____

9. $43 + 1 =$ _____

10. $29 + 1 =$ _____

11. $19 + 1 =$ _____

12. $26 + 1 =$ _____

13. $35 + 1 =$ _____

14. $67 + 1 =$ _____

15. $70 + 1 =$ _____

16. $54 + 1 =$ _____

17. $90 + 1 =$ _____

18. $39 + 1 =$ _____

19. $24 + 1 =$ _____

20. $89 + 1 =$ _____

21. $6 + 6 =$ _____

22. $3 + 3 =$ _____

23. $5 + 5 =$ _____

24. $7 + 7 =$ _____

25. $4 + 4 =$ _____

26. 2 groups of 6 = _____

26. 2 X 4 = _____

27. 2 X 3 = _____

28. 2 X 5 = _____

29. Half of 10 is_____

30. Half of 8 is _____

31. Half of 6 is _____

32. Half of 12 = _____

33. 5 + 2 (remember your fingers: 1 hand, 2 fingers) = _____

34. 5 + 1 = _____

35. 5 + 4 = _____

36. 5 + 3 = _____

37. 5 + 5 = _____

Lesson 13: Doubling 8, 9, and 10

Goals: **To learn to double the values of 8, 9, and 10**

New Teaching: We have doubled 2, 3, 4, 5, 6, and 7. Now let's double 8, 9, and 10. The easiest number to double is 10. Remember you have 10 fingers and 10 toes. How many is that all together? 20 total fingers and toes. You also know that 1 + 1 = 2. The same concept applies because 10 + 10 = 20. You simply carry the zeros along.

Example: 10 + 10 = 20, meaning 2 X 10 (2 groups of 10) = 20

Example: 8 + 8 = 16 and 2 X 8 = 16

Example: 9 + 9 = 18 and 2 X 9 = 18

Student Practice

Practice doubling, halves, and +1 addition.

1. Double 10 _____

2. Double 9 _____

3. Double 8 _____

4. 9 + 9 = _____

5. 2 X 9 = _____

6. 2 groups of 9 = _____

7. 8 + 8 = _____

8. 2 X 8 = _____

9. 2 groups of 8 = _____

10. 10 + 10 = _____

11. 2 groups of 10 = _____

12. 2 X 10 = _____

13. 8 + 8 = _____

14. A dozen eggs means there are _____ eggs.

15. Half of 12? _____

16. A half dozen cookies means there are _____ cookies.

17. Half of 4? _____

18. 8 + 1 = _____

19. 3 + 1 = _____

20. 85 + 1 = _____

21. 13 + 1 = _____

22. 27 + 1 = _____

23. 34 + 1 = _____

24. 79 + 1 = _____

25. 93 + 1 = _____

Notes:

_____

_____

_____

_____

_____

_____

_____

_____

_____

_____

_____

_____

Lesson 14: Counting with Dimes and Pennies

Goal: **To be able to count groups of dimes and pennies by switching between counting by 10's and counting by 1's.**

New Teaching: Counting by dimes and pennies is very similar to counting by nickels and pennies. This time, we have to count by 10's and then switch to counting by 1's.

Example: 10, 20, 30 SWITCH 31, 32, 33, 34

Example: 10, 20 SWITCH 21, 22, 23

Example:

10   20   30   40   SWITCH   41   42

Student Practice

Count by the pattern given (either 5's or 10's) and then switch to counting by 1's after the word switch.

1. 10, 20, 30 SWITCH 31, 32, _____, _____

2. 10, 20 SWITCH 21, 22, _____, _____, _____

3. 10, 20, _____, _____ SWITCH 41, 42, _____

4. 10, 20, 30, _____, _____ SWITCH 51, _____

5. 10, 20, 30, 40, 50, 60 SWITCH _____, _____, _____, _____

6. 10, _____, _____, _____, _____ SWITCH _____

7. 15, 25, 35, _____, _____ SWITCH 56, _____, 58, _____, _____

8.   _____ cents

9.   _____ cents

10. _____ cents

11. _____ cents

12. _____ cents

13. _____ cents

14. _____ cents

15. _____ cents

Notes:

_____

_____

_____

_____

_____

_____

_____

_____

_____

_____

_____

_____

_____

Lesson 15: Adding + 2 and + 3

Goals: **To learn how to add a number + 2 and + 3.**

New Teaching: When adding + 2 and + 3, the easiest way is to use a method called counting on. I like to put dots on my 2 and 3 to help.

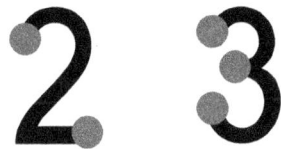

2 3

When adding two numbers follow these steps:

1. Circle the bigger number.

2. Say the bigger number.

3. Count on by touching the dots on the 2 or 3.

4. Write your answer.

So 6 + 2 would look this: ⑥ + 2

1. The 6 is bigger, so circle the six.

2. Say, "Six", then count on two using dots, "Seven, Eight."

3. Write the answer: 8.

Student Practice

Add numbers + 2 or + 3.

1. $4 + 2 =$ _____

2. $8 + 2 =$ _____

3. $2 + 6 =$ _____

4. $2 + 5 =$ _____

5. $5 + 3 =$ _____

6. $4 + 3 =$ _____

7. $3 + 7 =$ _____

8. $3 + 6 =$ _____

Student Practice

Practice adding + 1, + 2, + 3, and doubling.

1. $1 + 2 = $ _____

2. $5 + 2 = $ _____

3. $6 + 3 = $ _____

4. $7 + 2 = $ _____

5. $2 + 3 = $ _____

6. $7 + 3 = $ _____

7. $2 + 7 = $ _____

8. $8 + 3 = $ _____

9. $9 + 3 = $ _____

10. $9 + 2 = $ _____

11. $2 + 9 = $ _____

12. 3 + 6 = _____

13. 2 + 2 = _____

14. 4 + 4 = _____

15. 8 + 8 = _____

16. 5 + 5 = _____

17. 10 + 10 = _____

18. 2 groups of 4 = _____

19. 2 X 7 = _____

20. 9 + 9 = _____

21. 6 + 6 = _____

22. 5 + 5 = _____

23. 9 + 3 = _____

24. 10 + 3 = _____

Mixed Review

1. 59, 69, _____, _____, _____, _____

2. 25, 30, 35, _____, _____, _____, _____

3. 16, 21, 26, 31, _____, _____, _____, _____

4. 6 + 2 = _____

5. 2 + 2 = _____

6. 4 + 3 = _____

7. Double 5? _____

8. Half of 6? _____

9. 2 + 8 = _____

10. 5 + 3 = _____

11. 9 + 1 = _____

12. 3 + 1 = _____

13. 6 + 6 = _____

14. 2 X 9 = _____

15. 25, 50, _____, $1

16. _____ quarters are in 75 cents.

17. _____ cents

18. How many tallies? _____

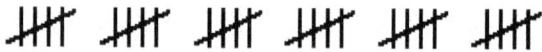

卌 卌 卌 卌 卌 卌

19. There are _____ eggs in half a dozen.

20. _____ cents

21. 5, 10, 15, _____, _____, _____ SWITCH 31, 32, _____, _____

22. What is the place value of 7 in 873? _____

23. What is in the hundreds place in 902? _____

Lesson 16: Addition Word Problems

Goal: **To learn how to apply addition word problems with known addition facts.**

Example: Kelly had 4 pieces of candy. Sara had 2 pieces of candy. How much candy in all?

Students may draw a picture (teach them to use circles to illustrate their items rather than drawing the "real" thing). Students should also write the number sentence for each:

$4 + 2 = 6$

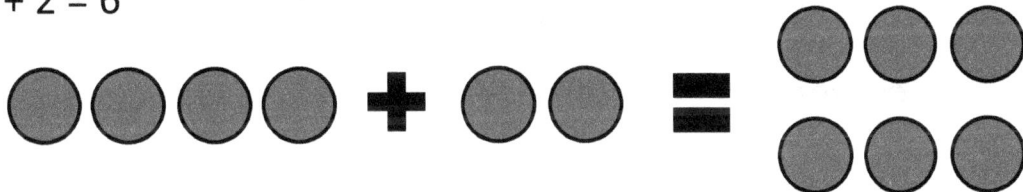

Word problems are challenging for some children. Teach them that they can look for key words to help them determine what type of problem it is. In our example, the words "in all" suggest that we are combining two groups and will be doing a + problem. As we expand in the types of problems we do, we will look for additional key words.

Student Practice

Use the addition strategies you have learned to solve these addition word problems.

1. Lilly has 2 apples and Jon has 2 apples. How many apples are there in all? _____

2. Fred has 5 stamps and Mike has 5 stamps. How many stamps are there in all? _____

3. Will has 8 books and Jen has 8 books. How many books are there in all? _____

4. Ken ate 4 cookies while Sam ate 1 cookie. How many cookies were eaten in all? _____

5. Jeff had 4 cats. He got 2 more cats. How many cats does he have in all? _____

6. Sara has 8 stickers and Tim gave her 2 more stickers. How many stickers does Sara have? _____

7. Sandy has 10 pens. Carl has 5 pens. How many pens are there in all? _____

8. Maria has 4 peanuts and Jeff has 4 peanuts, how many peanuts are there in all? _____

9. Lee has 10 M&M's. His mom has 6 M&M's. How many M&M's are there in all? _____

10. Corey has 1 bird. Tim has 2 birds. How many birds are there in all? _____

11. There is 1 cat in the red cage and 3 cats in the blue cage. How many cats are there in all? _____

12. Matthew has 5 marbles. Eric has 3 marbles. How many marbles are there in all? _____

13. Kathy has 2 markers. Katie has 4 markers. How many markers do they have in all? _____

Notes:

_____

_____

_____

_____

_____

_____

_____

_____

_____

_____

_____

_____

Lesson 17: Determining Operation for Word Problems + vs. –

<u>Goal</u>: **To determine if each situation is + or -.**

<u>New Teaching:</u> Students need to learn quickly to think and comprehend the story problem so that they can determine if the situation requires addition or subtraction. Initially, we will start with simple + and – situations and then build on key words to help us choose the right operation as we move forward.

For each problem, the student should state whether the problem is a situation where the two numbers are combined or added together or if the situation is a "take-away" problem that requires subtraction. Students can write the number sentences if they wish, but do not need to calculate the solutions at this time.

Student Practice

Tell whether this is an adding-on or a take-away problem.

1. There are 4 frogs on a log and 1 hopped off. How many frogs are left? _____

2. There are 10 friends and 3 more came. How many friends are there in all? _____

3. There were 8 slices of pizza. 2 slices were eaten. How many slices are left? _____

4. There are 9 stickers on a page. If 9 more stickers are placed on the page, how many stickers in all? _____

5. There are 10 dogs at the dog park. 4 dogs went home. How many dogs are left at the dog park? _____

6. There are 4 pencils are on a desk. 3 are taken away. How many pencils are left? _____

7. There are 8 marbles in a jar and 2 more are put in. How many marbles are there in all? _____

8. There were 5 balloons, but 3 popped. How many balloons are left? _____

9. There are 8 pairs of roller skates, but 1 pair broke. How many pairs of roller skates can be used? _____

10. There were 2 cookies in the cookie jar. Billy made 10 more. How many cookies are now in the cookie jar?

_____

11. There are 7 birds on a tree. A dog barked and scared 3 away. How many birds are left on the tree? _____

12. There are 4 kids at the park when 5 more joined them. How many kids are now at the park? _____

13. There are 8 lollipops in a bucket. If 4 are taken out, how many are left in the bucket? _____

Mixed Review

1. What is the place value of 8 in 983 (hundreds, tens, or ones)? _____

2. What is the place value of 6 in 639 (hundreds, tens, or ones)? _____

3. What is the place value of 2 in 72 (hundreds, tens, or ones)? _____

4. _____ cents

5. _____ cents

6. _____ cents

7. 10, 20, _____, _____, _____ **SWITCH** 51, _____, _____, _____

8. 2 x 5 = _____

9. Half of 10? _____

10. 4 + 2 = _____

11. 6 + 3 = _____

12. Double 8? _____

13. Half of 8? _____

14. Katie has 3 chocolate bars. Sarah has 4 chocolate bars. How many chocolate bars do they have together? _____

15. Lee had 8 stickers. His friend Dana gave him 3 stickers. How many stickers does Lee have now? _____

16. There were 9 people at Jacob's birthday party when it started, and 2 people showed up late. How many people are now at the birthday party? _____

Notes:

_____

_____

_____

_____

_____

_____

_____

_____

_____

_____

_____

_____

Lesson 18: Investigating the Hour Hand on the Clock

Goals: **To locate and name the hour hand on the clock.**

New Teaching: When learning to tell time, students will learn in two parts. We already introduced how to associate counting by 5's with hands on the clock. Now we want to teach students how to locate and name the hour hand on the clock. Things you need your student to know:

- The hour hand is the small hand.

- If the big hand is on the 12, choose the exact number the little hand is on. Otherwise, the hand will fall between 2 numbers. When choosing which number to use, you want the one that is the smaller of the two numbers. The only exception is when the hand falls between the 12 and the 1, you want to use the 12 (the one behind it). If you think about it, it is always "not quite" the next hour yet, so we want the one that is one hour less.

Example: State the number that the hour hand represents on the clock.

The big hand is not on the 12. It is on the 11. Therefore, when I look at the little hand, I see that it falls between the 11 and the 12. I choose the smaller of the two numbers: 11.

Example: State the number that the hour hand represents on the clock.

The big hand is on the 12. Therefore, when I look at the little hand, I choose the number it is on: 3

Example: State the number that the hour hand represents on the clock.

The big hand is not on the 12. It is on the 7. Therefore, when I look at the little hand, I see it falls between the 5 and the 6. I choose the smaller of the two numbers: 5.

Example: State the number that the hour hand represents on the clock.

The big hand is not on the 12. It is on the 9. Therefore, when I look at the little hand, I see it falls between 1 and 2. I select the smaller number: 1.

Student Practice

State the number that the hour hand (small hand) represents on the clock.

1.

The hour hand (small hand) represents _____ o'clock.

2.

The hour hand (small hand) represents _____ o'clock.

3.

The hour hand (small hand) represents _____ o'clock.

4. The hour hand (small hand) represents _____ o'clock.

5. The hour hand (small hand) represents _____ o'clock.

6. The hour hand (small hand) represents _____ o'clock.

7. The hour hand (small hand) represents _____ o'clock.

Mixed Review

1. How many fingers make up 6? 5 + _____

2. 13, 23, 33, _____, _____, _____, _____.

3. 26, _____, _____, _____, 66.

4. What comes after 54? _____

5. Double 10? _____

6. Half of 8? _____

7. 2 X 9 = _____

8. _____ cents

9. 20, 25, 30, _____, _____, _____ **SWITCH** 46, 47, _____, _____.

10. What is the place value of 9 in 69?   _____

11. What is the place value of 8 in 872?   _____

12. 6 + 3 = _____

13. 4 + 2 = _____

14. One little bird was sitting on a tree branch. 5 more birds landed on the tree branch. How many birds were there in all?

15. Katherine made 6 cupcakes for Michelle's birthday. Sarah made 3 cupcakes for Michelle's birthday. How many cupcakes are there all together? _____

State the number that the hour hand represents on the clock.

17.

The hour hand (small hand) represents _____ o'clock.

18.

The hour hand (small hand) represents _____ o'clock.

Notes:

_____

_____

_____

_____

_____

_____

_____

_____

_____

_____

_____

_____

Lesson 19: Finding Doubles + 1

Goal: **To identify and apply "doubles + 1" strategy.**

New Teaching: Students have learned all of their doubles. Now, we want students to be able to identify "doubles + 1". These are numbers that would be doubles except that one number is one more than the double.

Example: 5 + 5 is the double, so 5 + 6 is the "double + 1."

Example: 7 + 7 is the double, so 7 + 8 is the "double + 1."

Match the "double + 1" with its double.

5 + 6	3 + 3
3 + 4	7 + 7
8 + 9	8 + 8
7 + 8	5 + 5

<u>Example:</u>

4 + 5 is the "double + 1" for 4 +____

<u>Example:</u>

5 + 6 is the "double + 1" for 5 + ____

<u>Example:</u>

7 + 8 is the "double + 1" for 7 + ____

<u>Example:</u>

3 + 4 is the "double + 1" for 3 + ____

Student Practice

For each double, state it's double + 1.

1. 4 + 4 is the double, the double + 1 is ____ + ____

2. 6 + 6 is the double, the double + 1 is ____ + ____

3. 8 + 8 is the double, the double + 1 is ____ + ____

4. 1 + 1 is the double, the double + 1 is ____ + ____

5. 3 + 3 is the double, the double + 1 is ____ + ____

6. 5 + 5 is the double, the double + 1 is ____ + ____

7. 7 + 7 is the double, the double + 1 is ____ + ____

8. 2 + 2 is the double, the double + 1 is ____ + ____

9. 9 + 9 is the double, the double + 1 is ____ + ____

Now that the student is able to identify a double + 1 problem, show them how to apply the strategy when solving this type of problem.

<u>Example:</u> **4 + 4 = 8** so **4 + 5 = 9**

In this example, we see that 5 is one more than 4 so 4 + 5 is like the double 4 + 4, except it is one more. We know that 4 + 4 = 8, so if we add one more to 8, we get 9. Therefore, 4 + 5 = 9.

Student Practice

Use the doubles and double + 1 strategy to solve these problems. Do not use finger counting!

1. 3 + 3 = 6, so 3 + 4 = _____

2. 5 + 5 = 10, so 5 + 6 = _____

3. 2 + 2 = 4, so 2 + 3 = _____

4. 7 + 7 = 14, so 7 + 8 = _____

5. 9 + 9 = 18, so 9 + 10 = _____

6. 1 + 1 = 2, so 1 + 2 = _____

7. 6 + 6 = 12, so 6 + 7 = _____

8. 4 + 4 = 8, so 4 + 5 = _____

9. 8 + 8 = 16, so 8 + 9 = _____

10. 3 + 3 = _____ and 3 + 4 = _____

11. 7 + 7 = _____ and 7 + 8 = _____

12. $5 + 5 =$ _____ and $5 + 6 =$ _____

13. $8 + 8 =$ _____ and $8 + 9 =$ _____

14. $6 + 6 =$ _____ and $6 + 7 =$ _____

15. $4 + 4 =$ _____ and $4 + 5 =$ _____

16. $5 + 5 =$ _____

17. $5 + 6 =$ _____

18. $7 + 7 =$ _____

19. $7 + 8 =$ _____

20. $4 + 4 =$ _____

21. $4 + 5 =$ _____

22. $8 + 8 =$ _____

23. $8 + 9 =$ _____

24. $6 + 6 =$ _____

25. 6 + 7 = _____

26. 1 + 1 = _____

27. 1 + 2 = _____

28. 2 + 2 = _____

29. 2 + 3 = _____

30. 4 + 5 = _____

31. 7 + 8 = _____

32. 3 + 3 = _____

33. 3 + 4 = _____

34. 5 + 6 = _____

35. 9 + 9 = _____

36. 9 + 10 = _____

37. 8 + 9 = _____

Notes:

_____

_____

_____

_____

_____

_____

_____

_____

_____

_____

_____

_____

_____

Lesson 20: Finding Doubles – 1

Goal: **To notice and apply "doubles – 1."**

Just like your student was able to see the doubles + 1 pattern, we want them to see the "doubles – 1" pattern as well.

Example:

- 3 + 3 is the double.
- 3 + 4 is the double + 1.
- 3 + 2 is the double – 1.

Students should see all three situations and know that we add one to the double for "double + 1" and subtract one from the double in the "double – 1" case.

Example:

- 3 + 3 = 6   Double
- 3 + 4 = 7   Double + 1
- 3 + 2 = 5   Double – 1

Student Practice

Label each as a double, double + 1, or double – 1.

1. 4 + 4 _____

2. 4 + 3 _____

3. 4 + 5 _____

4. 8 + 9 _____

5. 5 + 6 _____

6. 1 + 2 _____

7. 6 + 6 _____

8. 6 + 5 _____

9. 2 + 2 _____

10. 3 + 2 _____

11. 9 + 8 _____

12. 2 + 1 _____

13. 7 + 8 _____

14. 7 + 7 _____

15. 5 + 4 _____

16. 9 + 10 _____

17. 3 + 4 _____

18. 4 + 4 _____

19. 7 + 6 _____

20. 2 + 3 _____

21. 3 + 3 _____

22. 8 + 8 _____

23. 9 + 9 _____

24. 8 + 7 _____

Student Practice

Fill in the blank using doubles and doubles – 1. Do not use finger counting!

1. 4 + 4 is the double so 4 + _____ is the **double – 1.**

2. 6 + 6 is the double so 6 + _____ is the **double – 1.**

3. 9 + 9 is the double so 9 + _____ is the **double – 1.**

4. 7 + 7 is the double so 7 + _____ is the **double – 1.**

5. 2 + 2 is the double so 2 + _____ is the **double – 1.**

6. 5 + 5 is the double so 5 + _____ is the **double – 1.**

7. 4 + 4 = 8 so 4 + 3 = _____

8. 8 + 8 = 16 so 8 + 7 = _____

9. 9 + 9 = 18 so 9 + 8 = _____

10. 5 + 5 = 10 so 5 + 4 = _____

11. 7 + 7 = 14 so 7 + 6 = _____

12. 3 + 3 = 6 so 3 + 2 = _____

13. 6 + 6 = 12 so 6 + 5 = _____

14. 9 + 9 = _____

15. 9 + 8 = _____

16. 7 + 7 = _____

17. 7 + 6 = _____

18. 5 + 5 = _____

19. 5 + 4 = _____

20. 6 + 6 = _____

21. 6 + 5 = _____

22. 8 + 8 = _____

23. 8 + 7 = _____

Student Practice

For each problem use doubles, doubles + 1, or doubles − 1 to find your answer. Do not use finger counting!

1.   4 + 5 = _____

2.   3 + 3 = _____

3.   8 + 7 = _____

4.   6 + 6 = _____

5.   5 + 6 = _____

6.   8 + 9 = _____

7.   4 + 4 = _____

8.   7 + 8 = _____

9.   9 + 9 = _____

10. 2 + 3 = _____

11. 6 + 7 = _____

12. 6 + 5 = _____

13. 2 + 1 = _____

14. 3 + 2 = _____

15. 5 + 5 = _____

16. 5 + 4 = _____

17. 8 + 8 = _____

18. 7 + 8 = _____

19. 9 + 9 = _____

20. 4 + 3 = _____

21. 7 + 7 = _____

22. 9 + 8 = _____

23. 7 + 6 = _____

24. 3 + 4 = _____

Mixed Addition Review

1. 6 + 1 = _____

2. 4 + 2 = _____

3. 6 + 3 = _____

4. 4 + 3 = _____

5. 10 = 5 + _____

6. 2 X 9 = _____

7. Half of 12? _____

8. Double 4? _____

9. 3 + 3 = _____

10. 5 + 6 = _____

11. 4 + 5 = _____

12. 9 + 8 = _____

Lesson 21: Adding + 9

Goal: **To add numbers + 9 by adding the original number to 10 and then take one less.**

New Teaching: The student should know how to add 4 + 10 to make 14. We now want them to be able to add 9 to a number by adding 10 and taking one less.

Example: 10 + 5 = 15 so 9 + 5 = 14

Practice them in both directions: 7 + 10 = 17 so 7 + 9 = 16

Sometimes, more than one addition strategy applies and students can pick their favorite approach.

Example: 8 + 10 = 18 so 8 + 9 = 17. They also learned 8 + 9 as a "doubles + 1" so they can use that as well. 8 + 8 = 16 so 8 + 9 = 17.

Our practice for this lesson will start with the + 9 strategy, but will mix in all the different addition strategies we have practiced and include a few X2 and take half to help the student remember.

Student Practice

1. 10 + 4 = 14 so 4 + 9 = ＿＿＿

2. 7 + 10 = 17 so 7 + 9 = ＿＿＿

3. 3 + 10 = 13 so 3 + 9 = ＿＿＿

4. 8 + 10 = 18 so 8 + 9 = ＿＿＿

5. 6 + 9 = ＿＿＿

6. 2 + 9 = ＿＿＿

7. 5 + 9 = ＿＿＿

8. 1 + 9 = ＿＿＿

9. 9 + 9 = ＿＿＿

10. Double 4? ＿＿＿

11. Double 7? ＿＿＿

12. 2 X 9 = ＿＿＿

13. Half of 4 = _____

14. 6 + 7 = _____

15. 10 + 5 = _____

16. 5 + 8 = _____

17. 10 + 4 = _____

18. 5 + 3 = _____

19. 3 + 9 = _____

20. 5 + 9 = _____

21. 9 + 9 = _____

22. 8 + 7 = _____

23. 10 + 1 = _____

24. 7 + 2 = _____

Notes:

_____

_____

_____

_____

_____

_____

_____

_____

_____

_____

_____

_____

Lesson 22: Practicing Addition Facts

<u>Goal:</u> **To practice all of the addition facts we have learned.**

It is time to work on the remaining addition facts. Students have a variety of strategies they can use. They are able count on (+ 1, + 2, or + 3). They know how to add 10's by making teens and can add 9's by making the teen and taking one less. They know how to add their doubles and know how to apply double + 1 as well as double − 1. They also learned how to visualize their fingers for facts that involve adding 5. Our goal is to get the child away from counting all facts and utilize these strategies as much as they can. Remind them to "visualize" the objects in their heads. We don't believe that students have to master something 100% to move on to future mathematics problems. We often find that working in the new area helps solidify the previous material they struggled with. Continue to encourage the student to use either memorization if they just "know" the answer or a strategy.

Below is a table that summarizes which facts have strategies we have used and the 10 facts that are left over.

	1	2	3	4	5	6	7	8	9	10
1	Double	Double + 1	+ 1	+ 1	Fingers	+ 1	+ 1	+ 1	+ 1	Teen
2	Double − 1	Double	Double + 1	+2	Fingers	+2	+2	+2	Add 10, then - 1	Teen
3	+ 1	Double − 1	Double	Double + 1	Fingers	+3	+3	+3	Add 10, then - 1	Teen
4	+ 1	+2	Double − 1	Double	Double + 1				Add 10, then - 1	Teen
5	+ 1	+2	+3	Double − 1	Double	Double + 1			Add 10, then - 1	Teen
6	+ 1	+2	+3		Double − 1	Double	Double + 1		Add 10, then - 1	Teen
7	+ 1	+2	+3		Fingers	Double − 1	Double	Double + 1	Add 10, then - 1	Teen
8	+ 1	+2	+3		Fingers		Double − 1	Double	Double + 1	Teen
9	+ 1	+2	+3	Add 10, then - 1	Fingers	Add 10, then - 1	Add 10, then - 1	Double − 1	Double	Teen
10	+ 1	Teen	Teen	Teen	Fingers	Teen	Teen	Teen	Teen	Double

You can see that there are only 10 facts that don't fall into one of the above strategies. Practice these additional facts with flashcards. You can also use the following to help:

6 + 4 = 10. This is a fact that makes 10 (all 10 facts should be memorized for help with subtraction).

6 + 8 = 14. Since 6 is one LESS than 7 and 8 is one MORE than 7, this fact will equal the same as 7 + 7. These are called equivalent facts. This can be used if the child is able to comprehend this more advanced form of patterning.

Student Practice

Use all of the addition strategies you have learned to solve these addition problems.

1. $1 + 4 =$ _____

2. $4 + 5 =$ _____

3. $2 + 7 =$ _____

4. $3 + 3 =$ _____

5. $1 + 2 =$ _____

6. $6 + 7 =$ _____

7. $9 + 9 =$ _____

8. $8 + 7 =$ _____

9. $5 + 7 =$ _____

10. $4 + 6 =$ _____

11. $8 + 5 =$ _____

12. $10 + 6 =$ _____

13. $1 + 9 =$ _____

14. $2 + 6 =$ _____

15. $3 + 9 =$ _____

16. $3 + 5 =$ _____

17. $6 + 9 =$ _____

18. $10 + 4 =$ _____

19. $10 + 7 =$ _____

20. $1 + 8 =$ _____

21. $9 + 4 =$ _____

22. $9 + 5 =$ _____

23. $10 + 3 =$ _____

24. $9 + 8 =$ _____

Notes:

_____

_____

_____

_____

_____

_____

_____

_____

_____

_____

_____

_____

_____

Lesson 23: Telling Time to 5 Minute Intervals

Goal: **To tell time to 5 minute interval.**

Students have learned how to locate the hour hand on the clock in a previous lesson. They have also learned how to count by fives based on the location of the minute hand. We will review those two concepts and then put them together, telling time to the five minute interval.

Student Practice

Tell the hour on the clock based on the position of the hour hand.

1.

The hour hand (small hand) represents _____ o'clock.

2.

The hour hand (small hand) represents _____ o'clock.

3.

The hour hand (small hand) represents _____ o'clock.

Student Practice

Counting by fives, look at the minute hand (big hand) and tell what value the minute hand refers to.

1.

The minute hand (big hand) represents _____ minutes.

2.

The minute hand (big hand) represents _____ minutes.

3.

The minute hand (big hand) represents _____ minutes.

Students will now have to locate both hands on the clock at the same time. First, they should locate the hour hand and tell the hour it references and then locate the minute hand and count by fives to find the minutes.

Lesson Dialogue: "Which is the hour hand?" The student responds, "The little hand."

"Which is the minute hand?" The student responds, "The big hand."

"Look to see if the minute hand is pointing to the 12. Is it?" The student answers, "No."

"If it is not right on the 12, how do we find the hour to use?" The student answers, "We look at the two numbers it falls between and choose the smaller one except when it falls between 12 and 1."

"Good, so what is the hour we want to use here?" The student answers, "Two."

"Correct. Now look at the minute hand. Counting by fives, what time is associated with the minute hand?" The student starts at 1 and counts by 5's. He gets to the four and says, "Twenty."

"Correct. The time is 2:20."

Student Practice

Use the clock to determine the number of minutes and the hour it is referring to. Then, determine the exact time.

1.

Minute hand tells us: _____ minutes.

Hour hand tells us it is _____ o'clock.

The time is _____:_____

2.

Minute hand tells us: _____ minutes.

Hour hand tells us it is _____ o'clock.

The time is _____:_____

3.

Minute hand tells us: _____ minutes.

Hour hand tells us it is _____ o'clock.

The time is _____:_____

4.

The time is _____:_____

5.

The time is _____:_____

6.

The time is _____:_____

7.  The time is _____:_____

8.  The time is _____:_____

9.  The time is _____:_____

10.  The time is _____:_____

11.  The time is _____:_____

12.  The time is _____:_____

Notes:

_____

_____

_____

_____

_____

_____

_____

_____

_____

_____

_____

_____

_____

Lesson 24: Counting Quarters and Pennies

<u>Goal</u>: **To count quarters and pennies together.**

<u>Lesson Dialogue:</u> "Do you remember how to count by 25's or how to count quarters?" The student answers, "25, 50, 75, $1."

"Yes. So how many quarters make a dollar?" The student says, "Four."

"Good. How many make 50 cents?" The student says, "Two."

"Now we are going to count by 25's, but switch to counting by ones when we see the pennies. Let's review how we did that for dimes and nickels first."

"If we had  then we would count 10, 20, 30, and now we switch to counting by ones: 31, 32, 33."

"You try one."  The student counts 10, 20, 30, 40 and then 41, 42.

"Good. Now let's do our nickels again."

"We would count by 5's first. 5, 10, 15, 20, 25. Now we see the pennies, so we switch and count by 1's. 26, 27, 28."

"You try."

The student counts: "5, 10, 15, 20, 25, 30, 31, 32."

"Now we will do the same with quarters."

"25, 50, 75, switch and count by ones, 76, 77, 78, 79."

"You try."

The student says "25, 50, 51, 52."

"Good job. Let's practice some more."

Student Practice

Find the value of each set of coins.

1. _____cents

2. _____ cents

3. _____ cents

4. _____ cents

5. _____ cents

6. _____ cents

Mixed Review

1.   _____ cents

2. $7 \times 2 =$ _____

3.   Half of $6 =$ _____

4.   Jimmy has 10 hats. He buys 3 more. How many hats does Jimmy have now? _____

5.

The hour hand shows us it is _____ o'clock

6.

The time is _____:_____

7. If the minute hand is on the 7, how many minutes does it

   represent? _____ minutes

8. 6 + 5 = _____

9. 3 + 6 = _____

10. 3 + 2 = _____

11. 6 + 6 = _____

12. 7 + 8 = _____

13. 10 + 3 = _____

14. Double 3? _____

15. 5 + 5 = _____

16. Half of 8? _____

17. 9 + 7 = _____

18. 2 X 9 = _____

Notes:

_____

_____

_____

_____

_____

_____

_____

_____

_____

_____

_____

_____

Lesson 25: Fact Families for + and –

Goals: **To understand fact families for + and –.**

Lesson Dialogue: "We are going to learn how addition (+) and subtraction or take-away (-) are related. Here are 5 apples. If I give you one more, how many do you have?" The student says, "6."

"Good. Let's say I take that one back again. How many do you have now?" The student says, "5."

"Excellent. We end up with what we originally had. In math, we would write this as: 5 + 1 = 6 and 6 – 1 = 5."

"Let's do another one. If I have 9 stickers and I give you 8 more, how many do you have?"

The student thinks. Have them write down 9 + 8. "What strategy is this?" The student says, "Doubles – 1."

"Correct. So we know that 9 + 9 is what?" The student says, "18."

"So, 9 + 8 must be what?" The student responds, "17."

"Right. So we know that 9 stickers plus 8 more stickers equals 17 stickers. But if I then take the 8 stickers back, I have the original 9 that I started with. Mathematically, that is $9 + 8 = 17$ and $17 - 8 = 9$."

"This is the start of a fact family. In a fact family, we have 3 numbers. There are two smaller ones and 1 bigger one. In this last one, 9 and 8 were the two smaller ones and 17 was the larger one. We know that $9 + 8$ and $8 + 9$ are the same thing. We get 17 either way. This gives us two of our facts in the fact family: $8 + 9 = 17$ and $9 + 8 = 17$. There is a special name for this called the commutative property. This is when we can switch the order of the numbers, but it doesn't change the answer."

Example: $3 + 6 = 9$ and $6 + 3 = 9$. The answer will always remain 9, no matter what order the two numbers being added are in.

"Our next two facts come from subtraction. I can take the biggest number and subtract one of the smaller numbers and the answer will be the other smaller number."

Example: $17 - 9 = 8$. I can also switch that to be $17 - 8 = 9$.

Take out 3 index cards and write the numbers 8, 9, and 17 on three cards. Then get three more cards and put a +, -, and an = on those.

Use the cards to arrange the fact family so the student visually sees how we are just finding the different combinations of these 3 numbers.

Take out 3 more cards and write 4, 5, and 9. Have the student manipulate the cards to make the four facts for this fact family. Have the student copy this down on paper.

Give the student 3 cards. "Make up an addition fact that you know and write the numbers on the cards." The student writes 2, 1, and 3 on the cards. You give them the +, -, and = cards and again they create the fact family. Have them write each fact down on paper.

"Now we will try it without the cards. Write a fact you know on paper." The student writes down a fact. "Now write the commutative addition fact." Remind the student what that means. "Good. Now set up the two subtraction facts."

"What if I give you $12 - 5 = 7$? Can you write the other subtraction fact?" The student writes, "$12 - 7 = 5$."

"Good. Now take the two smallest numbers and write the two addition facts." The student writes $5 + 7 = 12$ and $7 + 5 = 12$.

Student Practice

For each fact given, write the remaining 3 facts.

1. 2 + 3 = 5 ___ + ___ = ___ ; ___ - ___ = ___ ; ___ - ___ = ___

2. 6 + 3 = 9 ___ + ___ = ___ ; ___ - ___ = ___ ; ___ - ___ = ___

3. 9 – 4 = 5 ___ - ___ = ___ ; ___ + ___ = ___ ; ___ + ___ = ___

4. 8 – 7 = 1 ___ - ___ = ___ ; ___ + ___ = ___ ; ___ + ___ = ___

For each set of 3 numbers, write the fact family.

1. 3, 9, 12   ___ + ___ = ___   ___ + ___ = ___

               ___ - ___ = ___   ___ - ___ = ___

2. 4, 6, 10   ___ + ___ = ___   ___ + ___ = ___

               ___ - ___ = ___   ___ - ___ = ___

Write your own fact families.

1. __ + __ = __ ; __ + __ = __ ; __ - __ = __ ; __ - __ = __

2. __ + __ = __ ; __ + __ = __ ; __ - __ = __ ; __ - __ = __

3. __ + __ = __ ; __ + __ = __ ; __ - __ = __ ; __ - __ = __

Lesson 26: Subtraction – Counting Backwards

Goals: **To find 1 less and 2 less.**

Lesson Dialogue: "We are now going to begin subtracting. For small numbers like – 1 or – 2, we can just count backwards to find our answer."

Example: 8 – 1 would mean start at 8 and count backwards 1 time. 8, 7. So 8 – 1 = 7. If we have 6 – 2, count backwards twice. 6, 5, 4.   6 – 2 = 4.

"Now you try it: 5 – 1?" The student says "5, 4. 5 – 1 = 4."

"Good, how about 9 – 2?" The student says, "9, 8, 7. So 9 – 2 = 7."

Student Practice

Use the counting backwards strategy to answer these questions.

1. $9 - 1 =$ _____

2. $6 - 1 =$ _____

3. $4 - 1 =$ _____

4. $5 - 1 =$ _____

5. $7 - 1 =$ _____

6. $3 - 2 =$ _____

7. $6 - 2 =$ _____

8. $8 - 1 =$ _____

9. $3 - 2 =$ _____

10. $5 - 2 =$ _____

11. $9 - 2 =$ _____

12. $8 - 2 =$ _____

13. $4 - 2 =$ _____

14. $7 - 2 =$ _____

15. $5 + 2 =$ _____

16. $7 + 3 =$ _____

17. $4 + 4 =$ _____

18. $7 + 6 =$ _____

19. $10 + 4 =$ _____

20. $5 + 6 =$ _____

21. $2 + 9 =$ _____

22. $9 \times 2 =$ _____

23. $5 \times 2 =$ _____

24. $8 \times 2 =$ _____

Notes:

_____

_____

_____

_____

_____

_____

_____

_____

_____

_____

_____

_____

Lesson 27: Subtraction – Counting Up

Goal: To use a counting up strategy to subtract when the numbers you are subtracting are close to each other.

Lesson Dialogue: "In the last lesson, we used a counting down strategy to do subtraction. Now we want to talk about a counting up strategy. This strategy is used when the two numbers we are subtracting are fairly close to each other."

Example: "9 – 8. In this case if we were to count down, we would have to do a lot of counting. It would have looked like this:

9, 8, 7, 6, 5, 4, 3, 2, 1

This is too much to keep track of. However, we can note that 9 and 8 are numbers that are close to each other. In this case, we can count up rather than count backwards. To do this, we take the smaller number, 8, and see how many times we need to count up until we reach the other number. We say, '8 ➡ 9' and note that we only had to count up 1 notch (a "notch" is illustrated with an arrow here. Ultimately, we count the number of "notches" or arrows) until we got to the other number. The answer is the number of "notches" (arrows) you needed to count to get to your other number. In this case, I had to count up just one number to get from 8 to 9, so the answer is 1."

Example: 7 − 5. I note that 5 and 7 are pretty close to each other, so I can count 5 ➡ 6 ➡ 7. I noticed I counted twice (see the 2 arrows) to get from the 5 to the 7, so my answer is 2."

"You try: 8 − 6." The student says, "6 ➡ 7 ➡ 8. I counted up twice, therefore 8 − 6 = 2."

Have your child use their fingers when practicing. It may take some practice to get used to the difference between counting backwards and counting up. If the child is really struggling, just practice one type at a time — when they are solid, practice the other type, and finally mix the two. They need to know:

- When to use which strategy
- Counting backwards ends you up on the answer
- The answer for counting up is based on the number of times you need to count.

"Let's look at some problems, tell me in each case if it is better to count up or count backwards:"

9 – 1   (Count backwards)

7 – 2   (Count backwards)

7 – 6   (Count up)

9 – 3   (Count backwards)

8 – 7   (Count up)

6 – 4   (Count up)

12 – 1   (Count backwards)

15 – 13   (Count up)

6 – 2   (Count backwards)

9 – 8   (Count up)

7 – 5   (Count up)

8 – 6   (Count up)

"Good, now let's put it into practice. Make sure you decide if counting up or counting backwards is the best approach. Then do the problem."

Student Practice

For each problem, use either counting up or counting backwards. If you just "know" the answer without counting, feel free to write it.

1. $9 - 7 =$ _____

2. $8 - 1 =$ _____

3. $9 - 3 =$ _____

4. $5 - 4 =$ _____

5. $15 - 12 =$ _____

6. $10 - 1 =$ _____

7. $13 - 4 =$ _____

8. $16 - 14 =$ _____

9. $7 - 3 =$ _____

10. $9 - 5 =$ _____

11. $6 - 3 =$ _____

12. $10 - 3 =$ _____

13. $7 - 1 =$ _____

14. $8 - 2 =$ _____

15. $10 - 5 =$ _____

16. $9 - 6 =$ _____

17. $11 - 3 =$ _____

18. $10 - 7 =$ _____

19. $13 - 12 =$ _____

20. $15 - 5 =$ _____

21. $6 - 4 =$ _____

22. $9 - 4 =$ _____

23. $14 - 13 =$ _____

Notes:

_____

_____

_____

_____

_____

_____

_____

_____

_____

_____

_____

_____

_____

Lesson 28: Subtraction Word Problems

<u>Goal:</u> **To understand different types of questions that involve subtraction.**

<u>Lesson Dialogue:</u> "We are now going to learn about different types of problems that involve working with subtraction. We have done one type already. If I have 8 pieces of candy and I eat 2 pieces, how much candy do I have left? How would you do this problem?" The student says, "I would draw 8 circles, cross off 2, and count the remaining circles."

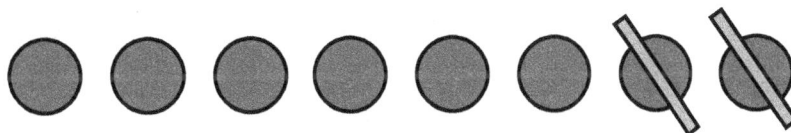

"Yes, using pictures is one way to solve the problem. Another way is to notice that the problem is subtraction. The key words that are often present in subtraction problems of this type are: how many are left?"

"We have the total number of candy (8), then subtract or take-away the 2 that got eaten: $8 - 2$. To solve this problem, would we count backwards or count up?" The student says, "Count backwards."

"Good. What do you get?" The student says, "8, 7, 6. Six."

"Yes, there are 6 pieces of candy left."

"Now we want to look at another type of subtraction word problem. This is when you are asked how many more."

Example: Carol had 10 trophies and her sister had 8 trophies. How many more trophies did Carol have?

"The key words, 'how many more,' tell us that this is a subtraction problem. We would do 10 – 8. Would we use count up or count backwards?" The student replies, "Count up."

"Good. What do you get?" The student says, "8 ➡ 9 ➡ 10. Two."

"Good. So that means Carol has 2 more trophies than her sister."

"Now you try one."

Example: Jill has 5 buckets of water. Jack has 1 bucket of water. How many more buckets of water does Jill have?"

"It is asking 'how many more' so I want to subtract 5 – 1. Since it is – 1, I want to count backwards: 5, 4. Jill has 4 more buckets of water than Jack."

"Excellent! So to review, we have two key phrases that clue us into subtraction: how many are left and how many more. Let's do one more type. In this type, the problem gives you a total number of something and one of the two parts that add to that total. You have to find the other part.

<u>Example:</u> There are a total of 10 balloons. Eight belong to Laura, the rest belong to Kyle. How many balloons does Kyle have?

"So, we have 10 total balloons. Some of the balloons are Laura's and some are Kyle's. We know that 8 are Laura's, so we need to figure out how many are left and those that are left must be Kyle's. This is the third type of subtraction problem. To solve, we take the total and subtract part to find the other part. In this case we would take 10 total balloons – 8 Laura balloons and the answer is the number of balloons that belong to Kyle. 10 – 8. To solve, we want to count up: 8 ➡ 9 ➡ 10. So the answer is 2. 8 of the 10 balloons are Laura's and the remaining 2 balloons are Kyle's."

"We will call this type of problem the part/total. Let's have you try one."

<u>Example:</u> There are 12 chairs in a room. Two chairs are empty, how many chairs are occupied?

"What is the total number of chairs?" The student says, "12."

"Right. The chairs are either empty or occupied. So we have a total to part relationship here. What would the subtraction sentence be?" The student replies, "12 – 2"

"Good, and will you count backwards or count on?" The student says, "Count backwards. 12, 11, 10. 10 chairs are occupied."

"Good. So we have 3 types of subtraction word problems:

- Part/total
- How many are left
- How many more

In the practice, you should first determine if the problem is a + problem or one of the 3 types of subtraction problems. Next, tell what type of subtraction problem it is or state addition. Then, write the number sentence, and finally find the answer."

Student Practice

1. Charlie buys 8 pieces of bubble gum. He gives away 6 pieces to his friends. How many pieces of bubble gum does Charlie have left?

    Type of problem: _____

    Number Sentence: _____

    Answer: _____

2. Michael has 10 marbles and Eddie has 7 marbles. How many more marbles does Michael have?

    Type of problem: _____

    Number Sentence: _____

    Answer: _____

3. Sarah and Bella share a piggy bank. There are 10 coins inside. If Bella puts 3 coins in the piggy bank, how many coins did Sarah put in?

Type of problem: _____

Number Sentence: _____

Answer: _____

4. Alice goes fishing with her dad. She catches 4 fish and her dad catches 5 fish. How many fish do they catch all together?

Type of problem: _____

Number Sentence: _____

Answer: _____

5. Together, Elizabeth and Joey have 9 stuffed animals. If 2 of them are Elizabeth's, how many belong to Joey?

Type of problem: _____

Number sentence: _____

Answer: _____

6. There were 10 ducks in the pond. Lucy feeds 8 of the ducks. How many ducks are not fed?

Type of problem: _____

Number sentence: _____

Answer: _____

7. There are 8 slices in a large pizza pie. If Marie eats 2 slices, how many slices are left?

Type of problem: _____

Number sentence: _____

Answer: _____

8. Chief and Bailey are dogs who love to play with toys. If there are 5 toys in the basket and 2 of them belong to Chief, how many toys belong to Bailey?

Type of problem: _____

Number sentence: _____

Answer: _____

9. Andy cooked 10 hamburgers. He then ate 2 and his family ate the rest. How many were left for his family to eat?

Type of problem: _____

Number sentence: _____

Answer: _____

10. Abby bought a box of 10 popsicles. If 6 of them are red popsicles and the rest are orange popsicles, how many popsicles are orange?

Type of problem: _____

Number sentence: _____

Answer: _____

11. It takes 9 hours to drive to New York. If a family drives 7 hours one day, how many more hours do they have to drive?

Type of problem: _____

Number sentence: _____

Answer: _____

12. Amelia has 8 stickers in her collection. If she gets 2 more, how many stickers does she have in all?

Type of problem: _____

Number sentence: _____

Answer: _____

Lesson 29: Subtraction – Build to 10

Goal: **To use the "build to 10" strategy for subtraction from teen numbers.**

New Teaching: In preparation to do additional subtraction problems, we need to work on the concept of building to 10. Since our number system is a system of 10's, we can take advantage of this by using knowledge such as build to 10. First, students need to be able to take the numbers 1 – 9 and know its partner that allows them to build to 10. The second part is being able to know a teen number and how far it is away from ten. We touched on this in our unit on building teens.

Lesson Dialogue: "Remember when we did the unit on building teens, you would be given 13 and asked $13 = 10 +$ _____. The easy way to do this was to just look at the ones place and your answer was there because 13 means 1 ten and 3 ones or $10 + 3$. In the build to 10 strategy, we use the same concept. How far away from 10 is 13? The answer is the one's place: 3. 13 is 3 away from 10. 18 is 8 away from 10. We call this build from 10.   Try these."

Student Practice

1. Build from 10 to 14. _____

2. Build from 10 to 12. _____

3. Build from 10 to 16. _____

4. Build from 10 to 19. _____

5. Build from 10 to 13. _____

6. Build from 10 to 11. _____

7. Build from 10 to 18. _____

8. Build from 10 to 15. _____

9. Build from 10 to 17. _____

Lesson Dialogue: "Next, we want to build to 10. How far from 10 is 5? Well, we know that 5 + 5 = 10, so 5 is 5 away from 10. How far is 6 from 10? Well, I need 4 more to get to 10. If you aren't able to do this quickly in your head yet, you can write down this easy chart and look at it to get your answers quickly. You write the numbers 1 – 5 and then underneath you write the numbers 5 – 9 backwards, so your chart looks like this:"

Building to 10

1	2	3	4	5
9	8	7	6	5

"Do you see the pattern from the chart? Cover the chart and use the pattern to write your own chart."

"Now, if we want to know how many 2 is from 10, we look below the 2 and see the answer is 8. If we want to know how far 6 is from 10, we look above the 6 and see the answer is 4. Try these using your chart."

Student Practice

1. Build from 7 to 10. _____

2. Build from 2 to 10. _____

3. Build from 5 to 10. _____

4. Build from 3 to 10. _____

5. Build from 9 to 10. _____

6. Build from 1 to 10. _____

7. Build from 4 to 10. _____

8. Build from 6 to 10. _____

9. Build from 8 to 10. _____

Lesson Dialogue: "Now we want to combine these steps into solving subtraction problems. This strategy is used when subtracting a teen number minus a single digit. I build each one to and from 10. In other words, I write down how far each number is from 10. Once I have those two numbers, I add them and that is my answer."

Subtraction Steps:

1) Build to 10

2) Build from 10

3) Add

Example: **13 – 8 =**

1. Build from 8 to 10: **2**

2. Build from 10 to 13: **3**

3. Add: 2 + 3 = **5**

Example: **15 – 7 =**

1. Build from 7 to 10: **3**

2. Build from 10 to 15: **5**

3. Add: 3 + 5 = **8**

Student Practice

Use the Build To/From 10 methods to solve these problems.

1. $14 - 6$

   Build to 10: _____

   Build from 10: _____

   Add: _____

2. $13 - 8$

   Build to 10: _____

   Build from 10: _____

   Add: _____

3. $12 - 4$

   Build to 10: _____

   Build from 10: _____

   Add: _____

4. $14 - 9$

   Build to 10: _____

   Build from 10: _____

   Add: _____

5. $15 - 8$

   Build to 10: _____

   Build from 10: _____

   Add: _____

6. $11 - 8$

   Build to 10: _____

   Build from 10: _____

   Add: _____

7. $15 - 9$

   Build to 10: _____

   Build from 10: _____

   Add: _____

8. 18 – 9

Build to 10: _____

Build from 10: _____

Add: _____

9. 16 – 8

Build to 10: _____

Build from 10: _____

Add: _____

10. 16 – 7

Build to 10: _____

Build from 10: _____

Add: _____

11. 14 – 7

Build to 10: _____

Build from 10: _____

Add: _____

12. 13 − 5

   Build to 10: _____

   Build from 10: _____

   Add: _____

13. 12 − 6

   Build to 10: _____

   Build from 10: _____

   Add: _____

14. 11 − 7

   Build to 10: _____

   Build from 10: _____

   Add: _____

15. 14 − 8

   Build to 10: _____

   Build from 10: _____

   Add: _____

Notes:

_____

_____

_____

_____

_____

_____

_____

_____

_____

_____

_____

_____

_____

Lesson 30: Word Problems

Goals: **To reinforce new subtraction situations and solve word problems.**

Lesson Dialogue: "In our previous lessons we have learned the 'build to 10' numbers in order to subtract single digit numbers from teen numbers. We also learned what to look for to determine if the problem is addition or subtraction:

- How many in all? (usually means +)
- How many are left? (usually means -)
- How many more? (usually means -)
- Total/part (usually means -)

Now let's try a few slightly more challenging problems that will require you to practice your 'build to 10' strategy when subtracting."

Student Practice

Use the Build To/From 10 methods to solve these problems.

1. Angie had 14 lollipops, but her sister took 5 of them. How many lollipops does Angie have left?

    Number sentence: _____

    Build to 10: _____

    Build from 10: _____

    Add: _____

2. There were 13 toy blocks in a box. David took 6 out to play with. How many blocks are left in the box?

    Number sentence: _____

    Build to 10: _____

    Build from 10: _____

    Add: _____

3. There are 15 M&M's in a jar. If Carrie ate 7 of the M&M's and Johnny ate the rest, how many M&M's did Johnny eat?

Number sentence: _____

Build to 10: _____

Build from 10: _____

Add: _____

4. There are 13 straws in a jar. If 5 are taken out to be used, how many straws are left?

Number sentence: _____

Build to 10: _____

Build from 10: _____

Add: _____

5. There are 14 buttons in a box. If 9 are taken out, how many buttons are left in the box?

   Number Sentence: _____

   Answer: _____

6. There are 13 apples in the tree. If 9 are picked, how many apples remain on the tree?

   Number Sentence: _____

   Answer: _____

7. There are 11 baseballs in a box. If 6 of them are removed to play with, how many baseballs are left in the box?

   Number Sentence: _____

   Answer: _____

8. Gabe placed 10 bricks on top of each other to make a pyramid. If 7 are removed to make a smaller pyramid, how many bricks are in the new pyramid?

Number Sentence: _____

Answer: _____

9. There are 12 buttons in a box. If 5 are taken out, how many buttons are left in the box?

Number Sentence: _____

Answer: _____

10. Mitchell and Jaime put 10 toy boats in the pool. If 4 belong to Mitchell, how many belong to Jaime?

Number Sentence: _____

Answer: _____

Mixed Word Problem Review

Determine what type of word problem (+ or any of the 3 subtraction problems), write a number sentence, and solve.

1. Erin captured 10 fireflies one night and 4 more the next night. How many fireflies did she capture altogether?

    Type of problem: _____

    Number Sentence: _____

    Answer: _____

2. There were 16 cookies in the cookie jar. If 7 were eaten in one day, how many cookies remain?

    Type of problem: _____

    Number Sentence: _____

    Answer: _____

3. Leah and Jackie were jumping rope. If Leah jumped 14 times and Jackie jumped 5 times, how many more times did Leah jump than Jackie?

    Type of problem: _____

    Number Sentence: _____

    Answer: _____

4. There are 5 dogs at the park on Tuesday. On Wednesday, 2 more joined them. How many dogs were at the park in all on Wednesday?

    Type of problem: _____

    Number Sentence: _____

    Answer: _____

5. Two baseball teams are playing a game. The Tigers have 5 players while the Cubs have 11 players. How many more players do the Cubs have?

Type of problem: _____

Number Sentence: _____

Answer: _____

6. Cody has 6 stamps. He bought 6 more stamps. How many stamps does Cody have now?

Type of problem: _____

Number Sentence: _____

Answer: _____

7. Matthew had 18 stickers and gave 9 of them away to his friends. How many stickers does Matthew have left?

   Type of problem: _____

   Number Sentence: _____

   Answer: _____

8. There are 12 students in a class. If 3 students were absent on Friday, how many students were in the classroom that day?

   Type of problem: _____

   Number Sentence: _____

   Answer: _____

9. Jerry scored 9 baskets in a basketball game. His brother Tom scored 8 baskets. How many more baskets did Jerry score?

Type of problem: _____

Number Sentence: _____

Answer: _____

10. Leo had 12 toy soldiers. If he lost 3 of them, how many soldiers does Leo have left?

Type of problem: _____

Number Sentence: _____

Answer: _____

11. Marissa has a book with 9 pressed flowers. On a walk today, she found 3 more flowers to put in her book. How many flowers does Marissa have now?

Type of problem: _____

Number sentence: _____

Answer: _____

12. Monkey Steve loves to eat bananas. If he started off with 15 bananas and ate 8 of them, how many bananas does Monkey Steve have left?

Type of problem: _____

Number sentence: _____

Answer: _____

Notes:

_____

_____

_____

_____

_____

_____

_____

_____

_____

_____

_____

_____

_____

Lesson 31: Basic Arithmetic Practice

Goal: **To reinforce basic arithmetic skills, including addition, subtraction, multiplying by 2, and taking half.**

This lesson serves as a practice of basic arithmetic skills taught so far. We want to reinforce memorization, addition and subtraction strategies, what it means to multiply by 2, and how to take half of a given number.

Student Practice

Use all addition, subtraction, doubling, and taking half strategies to solve these problems.

1. $3 + 5 = $ _____

2. $5 + 4 = $ _____

3. $10 + 5 = $ _____

4. $10 + 7 = $ _____

5. $10 + 6 = $ _____

6. $3 + 4 = $ _____

7. $7 + 3 = $ _____

8. $2 + 8 = $ _____

9. $2 + 3 = $ _____

10. $2 + 6 = $ _____

11. $9 + 5 = $ _____

12. 9 + 3 = _____

13. 8+ 9 = _____

14. 9 + 9 = _____

15. 4 + 4 = _____

16. 5 X 2 = _____

17. 9 X 2 = _____

18. Half of 12? _____

19. Half of 8? _____

20. 3 − 2 = _____

21. 6 − 1 = _____

22. 8 − 2 = _____

23. 15 − 6 = _____

24. 17 − 8 = _____

Notes:

_____

_____

_____

_____

_____

_____

_____

_____

_____

_____

_____

_____

Lesson 32: Counting Dimes and Nickels

Goal: **To count dimes and nickels together.**

Lesson Dialogue: "We have learned how to count by 10's for dimes and then switch to counting by 1's for pennies. Now we need to count by 10's for dimes and switch to counting by 5's for nickels. Let's see how this would work:

10, 20, 30 SWITCH (count by 5's) 35, 40, 45, 50

Let's start by practicing just counting by 5's from a starting pointing other than 5."

Student Practice

Count by 5's from the point given.

1. 40, _____, _____, _____

2. 10, _____, _____, _____, _____

3. 15, _____, _____, _____, _____, _____

4. 25, _____, _____, _____, _____, _____

5. 55, _____, _____, _____

6. 75, _____, _____, _____, _____

7. 60, _____, _____, _____, _____, _____

8. _____, 35, 40, _____, _____, _____

9. _____, 55, _____, _____, 70, _____

10. 50, _____, _____, _____

11. 30, _____, _____, _____, _____

"Now we will count by 10's and then switch to counting by 5's."

Start of by counting by tens. 10, 20, 30, 40. Now switch and count by 5's starting at 40: 45, 50, 55, 60."

Take out some dimes and nickels and have the students use the dimes and nickels as counters. The most challenging part is getting the child used to the switch where they have to go from counting by 10's to counting by 5's.

Student Practice

Count by 10's and then switch to count by 5's.

1. 10, 20, 30, 40, 50 SWITCH ____, ____, ____

2. 5, 15, 25, ____, ____ SWITCH ____, ____, ____

3. 55, 65, ____, ____ SWITCH ____, ____, ____

4. 60 , ____, ____ SWITCH ____, ____, ____

5. 30, 40, ____, ____, ____ SWITCH ____, ____

6. 65, ____, ____ SWITCH ____, ____, ____

7. ____, 30, 40, ____, ____ SWITCH ____, ____

8. 65, ____, ____ SWITCH ____

9. 25, ____, ____ SWITCH ____, ____, ____

Lesson Dialogue: "Now we will do the same thing but with coins this time. Remember to switch when we change from dimes to nickels."

Example:

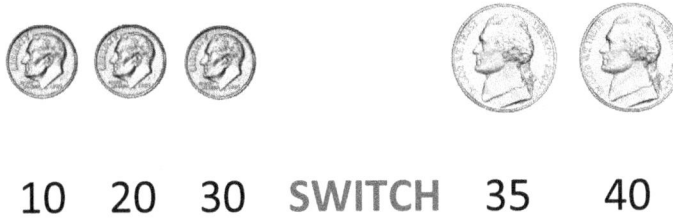

10   20   30   SWITCH   35   40

Example:

10   20   30   40   50   SWITCH   55   60

Example:

10   20   SWITCH   25   30   35   40

Student Practice

Count the dimes (by 10's) and then switch to count the nickels (by 5's).

1.

_____ _____ _____ SWITCH _____ _____ _____

2.

_____ _____ _____ _____ _____ SWITCH _____ _____

3. _____ cents

4. _____ cents

5. _____ cents

Lesson 33: Counting Three Different Coins

Goal: **To introduce counting 3 types of coins at one time.**

Lesson Dialogue: "Let's quickly review counting 2 coins at one time before we move on to counting 3 coins."

Student Practice

Find the correct number of cents for each problem.

1. ____ cents

2. ____ cents

3. ____ cents

4. ____ cents

5. ____ cents

<u>Lesson Dialogue:</u> "Now we have to learn to switch twice when doing our counting. Let's start with 3 dimes, 3 nickels, and 3 pennies."

<u>Example:</u>

"Our problem would look like this:

___, ___, ___, SWITCH ___, ___, ___, SWITCH ___, ___, ___

First we count by 10's: 10, 20, 30. Then switch and count by 5's: 35, 40, 45. Switch again and count by ones: 46, 47, 48."

<u>10</u>, <u>20</u>, <u>30</u> SWITCH <u>35</u>, <u>40</u>, <u>45</u> SWITCH <u>46</u>, <u>47</u>, <u>48</u>.

"Let's do another one."

<u>Example:</u>

"Count how many of each coin there are." The student says, "There is 1 dime, 4 nickels, and 4 pennies."

"Good. Here is what the setup will look like."

___ SWITCH ___, ___, ___, ___ SWITCH ___, ___, ___, ___

"First, we'll count the dimes by counting by 10's." The student responds, "It's only 10 because there's only one dime."

"Now switch to nickels and count by 5's." The student says, "15, 20, 25, 30."

"Switch one more time and count the pennies by counting by 1's." The student counts, "31, 32, 33, 34. There are 34 cents."

<u>10</u> SWITCH <u>15</u>, <u>20</u>, <u>25</u>, <u>30</u> SWITCH <u>31</u>, <u>32</u>, <u>33</u>, <u>34</u>

"There is one more I would like for you to try."

Example:

"Count each of the coins." The student responds, "There are 5 dimes, 2 nickels, and 2 pennies."

___, ___, ___, ___, ___ SWITCH ___, ___ SWITCH ___, ___

"Start by counting the dimes." The student says, "10, 20, 30, 40, 50."

"Good. Now switch to counting by 5's to count the nickels." The student counts, "55, 60"

"Excellent. Now count the pennies by counting by 1's." The student says, "61, 62. There are 62 cents."

<u>10</u>, <u>20</u>, <u>30</u>, <u>40</u>, <u>50</u> SWITCH <u>55</u>, <u>60</u>, SWITCH <u>61</u>, <u>62</u>.

"Right. Now let's try some on your own."

Student Practice

Find the correct number of cents using the strategies taught.

1.

_____ _____ _____ SWITCH _____ SWITCH _____ _____

2.

_____ _____ SWITCH _____ SWITCH _____ _____ _____

3.

_____ _____ _____ SWITCH _____ SWITCH _____

4.

_____ SWITCH _____ _____ _____ _____ SWITCH _____

5.

_____ _____ SWITCH _____ SWITCH _____ _____

6.

_____ _____ SWITCH _____ _____ _____ SWITCH _____

7.  _____ cents

8.  _____ cents

9.  _____ cents

Notes:

_____

_____

_____

_____

_____

_____

_____

_____

_____

_____

_____

_____

Lesson 34: Introduction to Making Exchanges

<u>Goal:</u> **To learn to make exchanges.**

<u>Lesson Dialogue:</u> "We use a base 10 system. This means that every time I get 10 units, I can exchange it for one of the next unit. Think of it like money. If someone gave you one $10.00 bill, it would be the same as getting ten $1.00 bills."

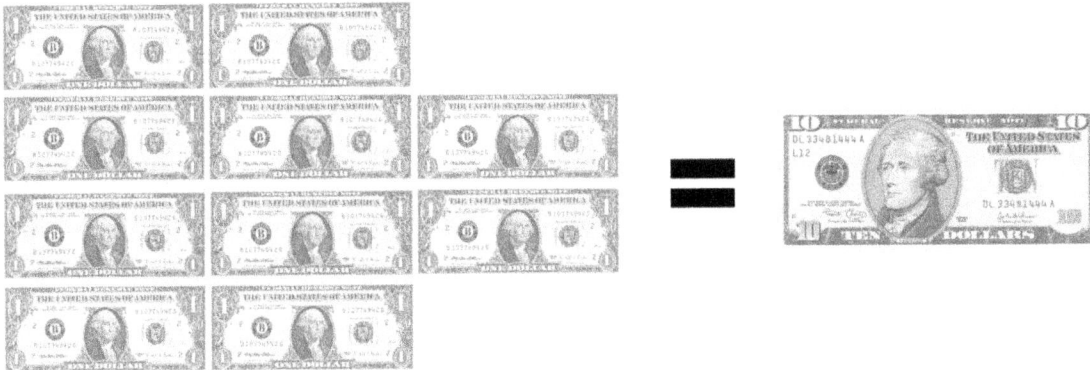

"10 pennies also equals 1 dime."

"When adding, we can exchange 10 ones for 1 ten. We can also exchange 10 tens for 1 hundred."

"The number 32 can mean either 32 ones (like 32 pennies) or it can mean 3 tens and 2 ones (like 3 dimes and 2 pennies)."

<u>Example:</u> 43 = 4 dimes and 3 pennies

Student Practice

1. 43 dollars = _____ tens and _____ ones

2. 25 cents = _____ dimes and _____ pennies

3. 20 dollars = _____ tens OR _____ ones

4. 15 cents = _____ dimes and _____ pennies

5. 81 dollars = _____ tens and _____ ones

6. 69 cents = _____ dimes and _____ pennies

7. 73 dollars = _____ tens and _____ ones

8. 98 dollars = _____ tens and _____ ones

9. 92 cents = _____ dimes and _____ pennies

10. 53 cents = _____ dimes and _____ pennies

11. 34 cents = _____ dimes and _____ pennies

12. 62 dollars = _____ tens and _____ ones

Mixed Review

1. 10, 20, 30, ____, ____, ____, ____, ____, ____, ____

2. 5, 10, 15, 20, ____, ____, ____, ____, ____, ____, ____,

3. 25, 30, 35, ____, ____, ____ SWITCH TO ONES ____, ____

4.    The time is ____:____

5. If there are 9 blue fish and 2 red fish, how many fish are there total? ____

6. 10 + 1 = ____

7. 68 + 1 = ____

8. 5 + 2 = ____

9. 93 + 2 = ____

10. 10 + 3 = _____

11. 26 + 3 = _____

12. Double 2? _____

13. Half of 10? _____

14. _____ cents

15. _____ cents

16. _____ cents

17. _____ cents

18. _____ cents

19. _____ cents

20. 9 + 3 = _____

21. 9 + 8 = _____

22. 9 + 9 = _____

23. 14 − 8 = _____

24. 14 − 2 = _____

25. 16 − 8 = _____

26. 5 − 1 = _____

27. 12 − 9 = _____

www.ingramcontent.com/pod-product-compliance
Lightning Source LLC
Chambersburg PA
CBHW051344200326
41521CB00014B/2475